Oxford International Primary Geography

Workbook

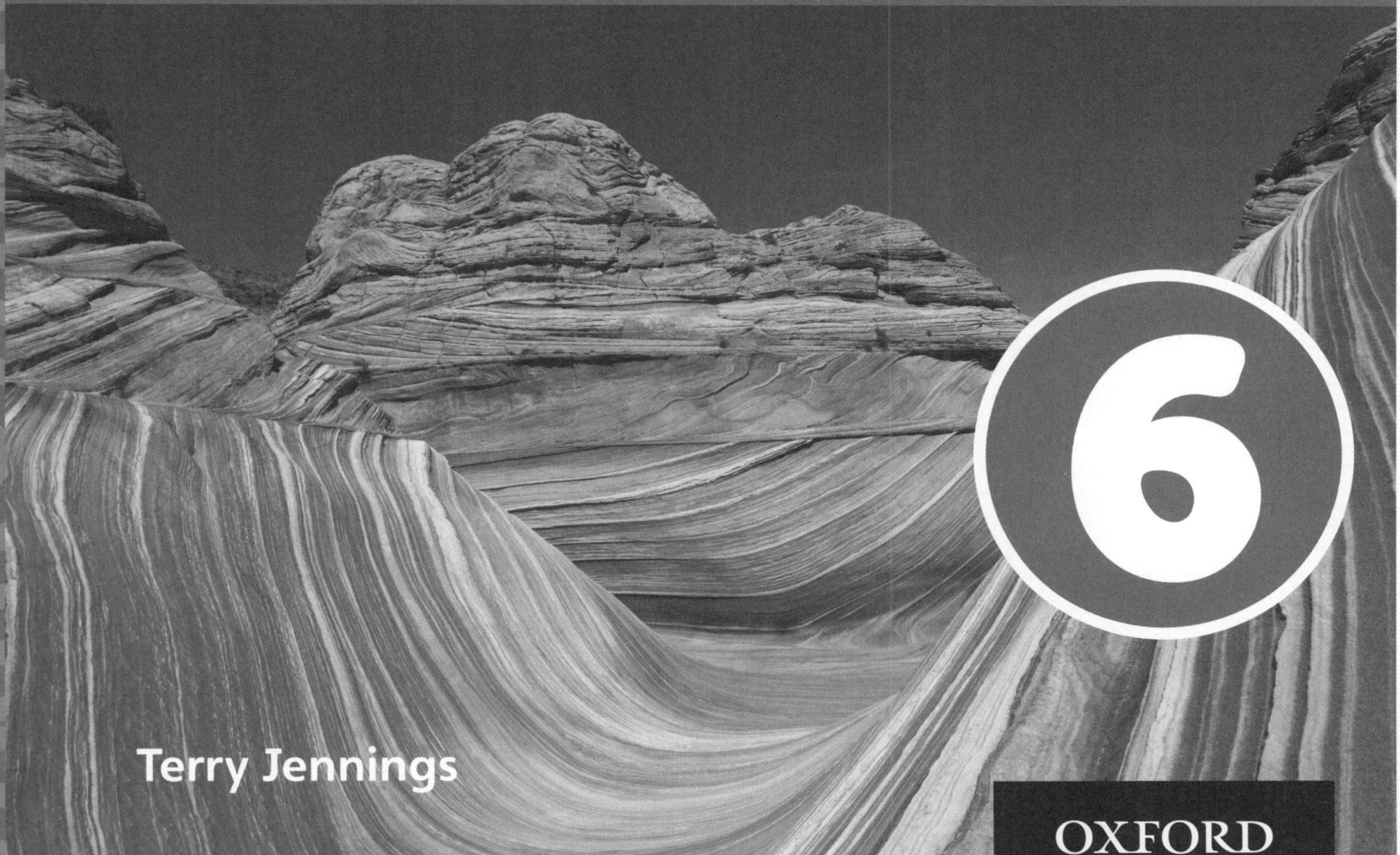

6

Terry Jennings

OXFORD
UNIVERSITY PRESS

Great Clarendon Street, Oxford, OX2 6DP, United Kingdom

Oxford University Press is a department of the University of Oxford. It furthers the University's objective of excellence in research, scholarship, and education by publishing worldwide. Oxford is a registered trade mark of Oxford University Press in the UK and in certain other countries

British Library Cataloguing in Publication Data
Data available

978-019-831-014-3

20

The manufacturing process conforms to the environmental regulations of the country of origin.

Printed and bound in Great Britain by Bell and Bain Ltd, Glasgow

Acknowledgements
The publishers would like to thank the following for permissions to use their photographs:

Cover photo: Fotofeeling/Westend61/Corbis

Although we have made every effort to trace and contact all copyright holders before publication this has not been possible in all cases. If notified, the publisher will rectify any errors or omissions at the earliest opportunity.

Links to third party websites are provided by Oxford in good faith and for information only. Oxford disclaims any responsibility for the materials contained in any third party website referenced in this work.

The manufacturer's authorised representative in the EU for product safety is Oxford University Press España S.A. of El Parque Empresarial San Fernando de Henares, Avenida de Castilla, 2 – 28830 Madrid (www.oup.es/en or product.safety@oup.com). OUP España S.A. also acts as importer into Spain of products made by the manufacturer.

Contents

Where does water come from?

The water cycle

Here is a simplified diagram of the water cycle. Use words from the word box to write the correct labels on the diagram.

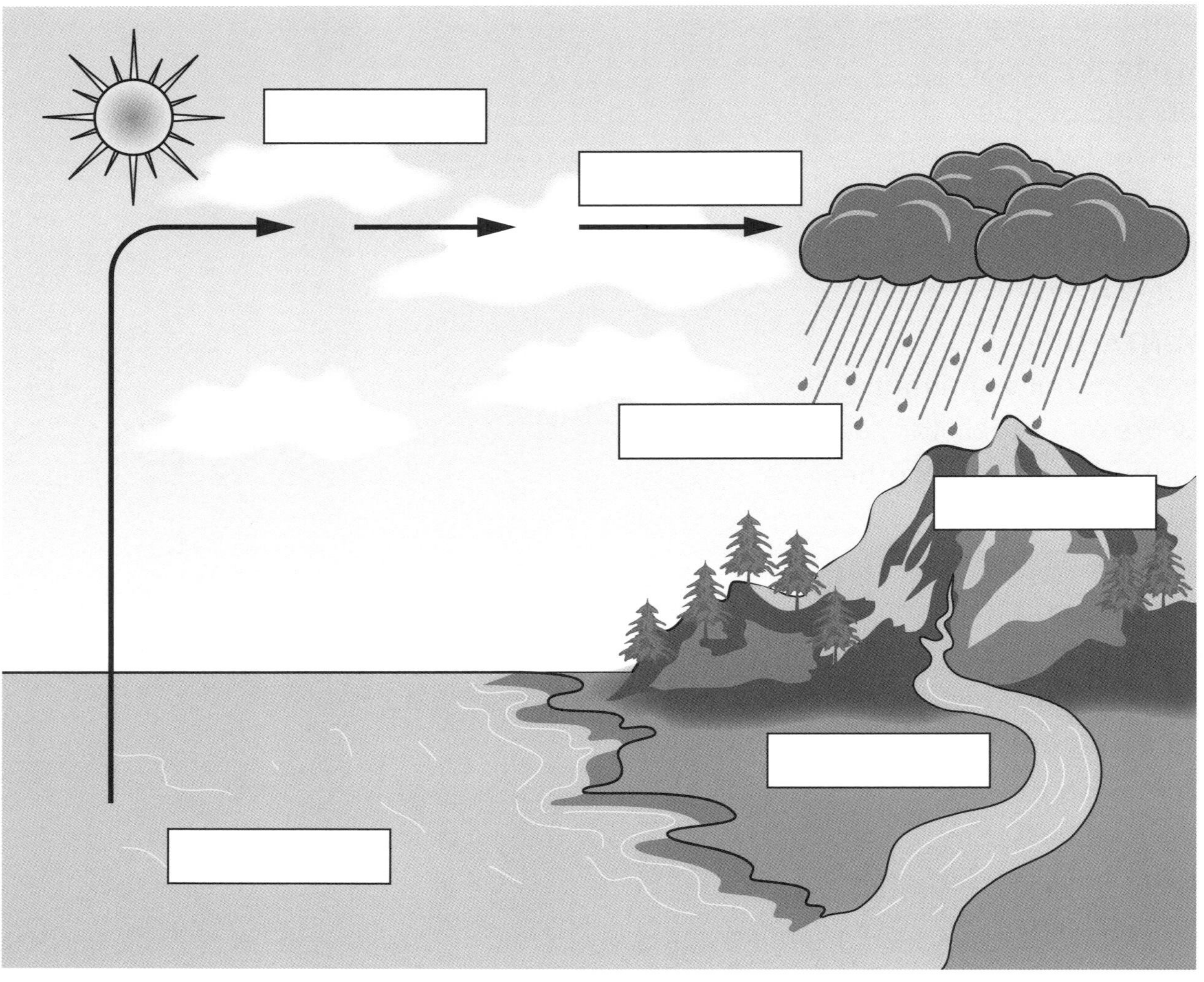

Sun	**river**	**cloud**
rain	**mountain**	**ocean**

- Where is most of the water in the water cycle stored? _______________

- What is the name of the process that changes water in the oceans and

  seas into water vapour in the air? _______________

- What is the name of the process that changes water vapour in the air

  into tiny water droplets in clouds? _______________

A model of the water cycle

Make a model of the water cycle.

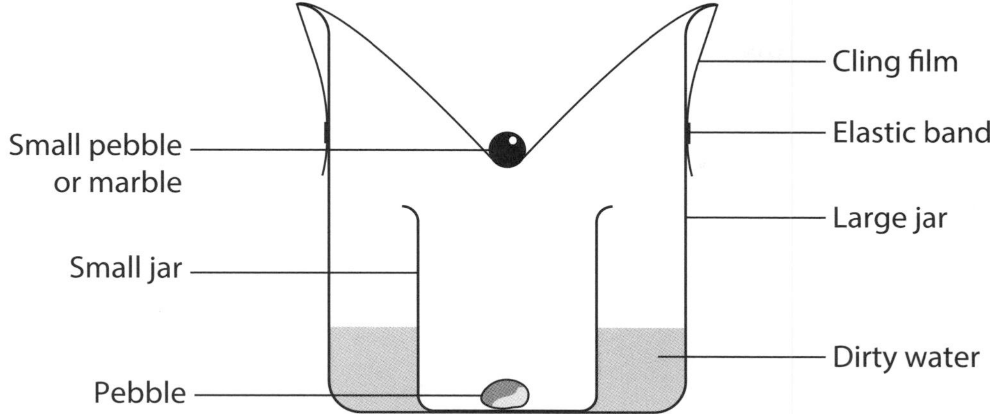

- Stir a little soil or food colouring into some water in a jug.

- Stand a small jar in the centre of a large clear plastic jar, and place a small, clean pebble inside it (to stop it floating).

- Carefully, without splashing, pour some of the dirty water into the large jar to a depth of a few centimetres.

- Cover the jar loosely with cling film. Press the centre of the cling film down so that it forms a cone shape. You could put a small pebble or a marble in the bottom of the cone to hold it down.

- Fasten the cling film in place with an elastic band.

- Stand the jar in a sunny place. Look at it carefully over the next day or so.

- What has happened? Explain what you see, using the words from the word box.

heat	Sun	evaporates	condenses	water droplets

____

____

____

____

Where rivers begin

The course of a river

As soon as rain falls or snow melts on high ground, the water starts to flow downhill. It tries to find the easiest way to the sea. The picture below shows the course of a river from the mountains to the sea.

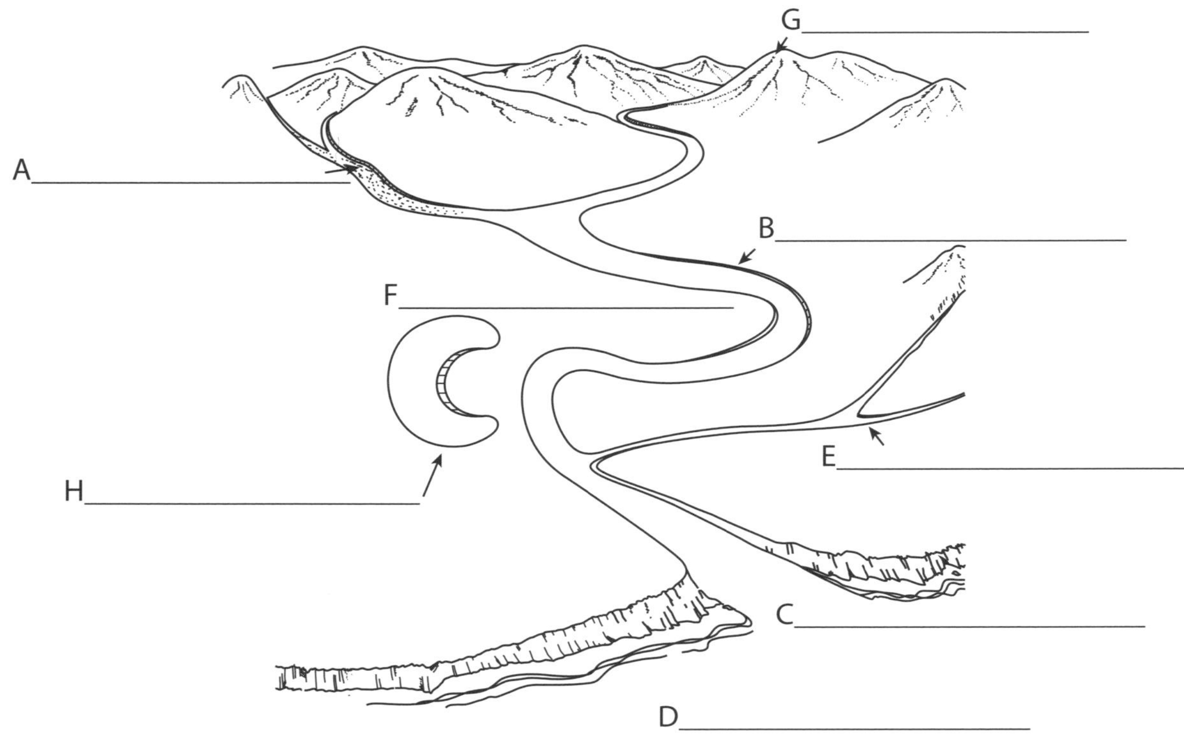

You could read pages 8 and 9 of the Student Book to help you answer the following questions.

- What are the names of the features labelled **A, B, C, D, E, F, G and H**?
  Use the words from the word box to write the correct label for each feature.

oxbow lake	**waterfall**	**tributary**	**mouth**
meander	**source**	**sea**	**floodplain**

- Use the picture to help you to fill in the gaps in these sentences.

The place where a river begins is called its ________________.

A large bend in a river is called a ________________.

A ________________ is a place where a river flows over a steep drop into a pool below.

River valleys

The speed of flow of a river and the shape and size of its valley change between the river's source and the sea. The pictures below show three of the stages in the life of a river.

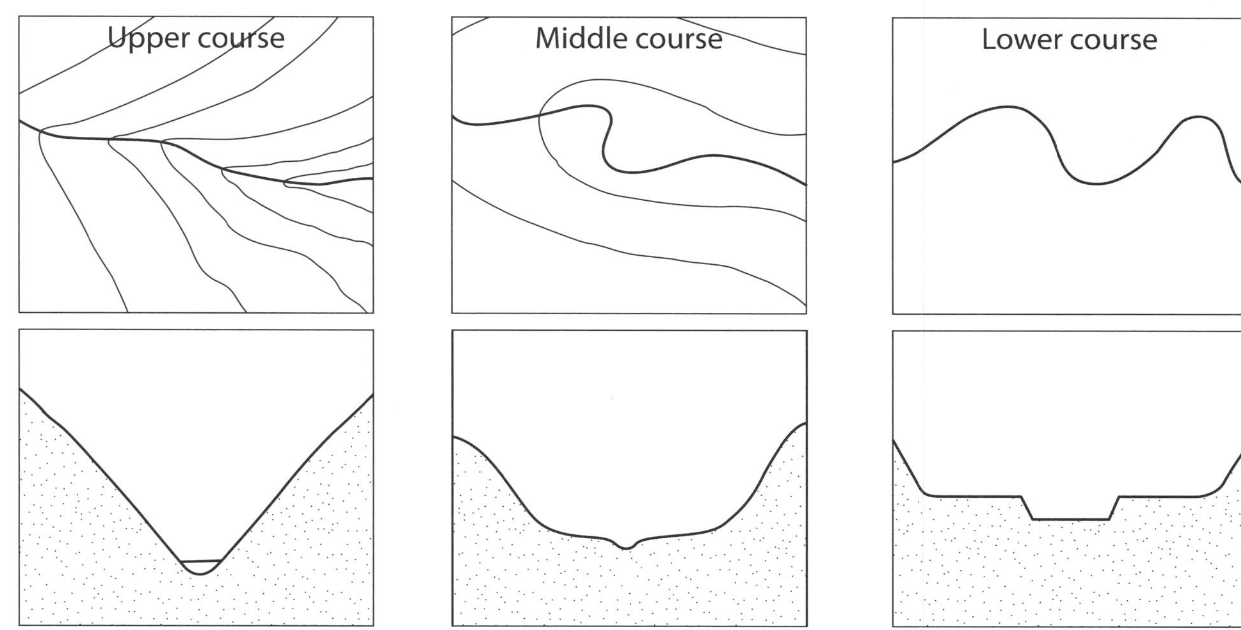

Look at the diagrams. For each diagram, write a sentence describing the valley and the river. Include words from the word box.

meanders	**fast**	**estuary**	**floodplain**
narrow	**flat**	**steep**	**flatter**

1 The upper course of the river is in the mountains. The river flows

_________________ and its valley floor is _________________

although the valley has _________________ sides.

2 In its middle course the river valley is wider. The valley bottom is

_________________. The river often forms _________________ across its valley.

3 In its lower course, the valley is very _________________. It is called a

_________________ because the river may overflow over it after

heavy rain. The river has large meanders and near its mouth the river may

form an _________________.

Down to the coast

Using estuaries

An estuary is the wide mouth of a river where it enters the sea. The level of the water rises with each high tide and falls at low tide. At low tide, estuaries have large areas of wet mud and shallow water. Large flocks of wading birds and wildfowl feed on estuaries at low tide.

Look at the picture of an estuary. How is it being used by people?

____

____

____

____

____

____

How might the human activities on the banks of the estuary harm the wading birds and wildfowl that feed in it?

____

____

____

Use maps and reference books or the Internet. Find the names of four large estuaries that are important wildlife sites and are also used by industry.

1 _______________________________   2 _______________________________

3 _______________________________   4 _______________________________

Settlements near estuaries

Use an atlas to find ten large river estuaries. Which rivers flow into them? Which large towns and cities are near these estuaries? Write what you discover.

Continent	Country	River	Large towns and cities near the estuary

Rivers and people

Continents and rivers

Look at the map of the world below. It shows the longest river on each continent.

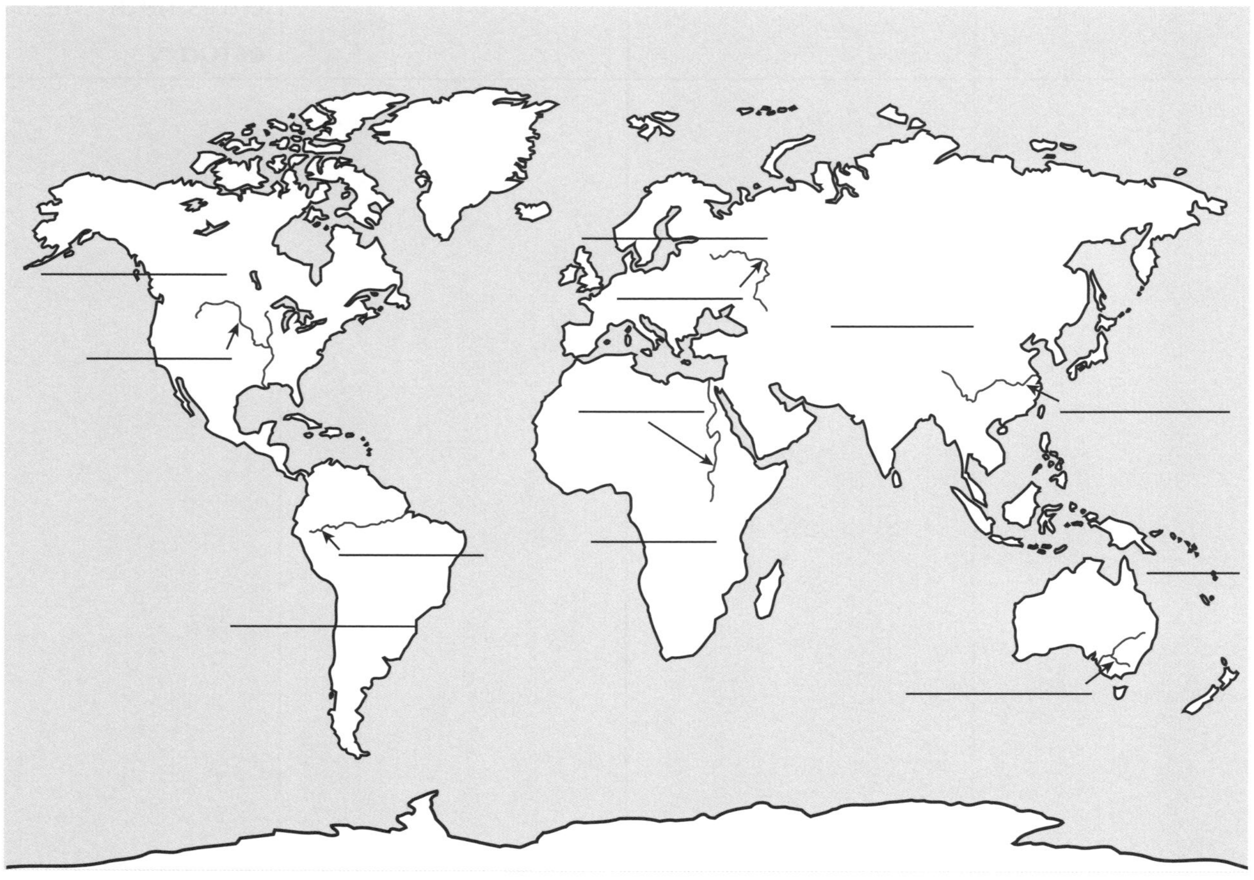

Continent	Rivers
Africa	Amazon
Asia	Mississippi-Missouri
Europe	Murray-Darling
North America	Nile
Oceania	Volga
South America	Yangtze or Chang Jiang

Use an atlas or the Internet to help you to label each of the rivers and to find out the length of each one. Write the length next to its name.

What are the two longest rivers in your country?

1 _________________________________ 2 _________________________________

Rivers and settlements

Early villages were often built near rivers and streams. Many of these villages later grew into towns and cities.

- Read the descriptions below and label each picture with the correct letter: **A, B, C** or **D**.

A The inside of a large bend, or meander, in the river provides protection from enemies.

B A small hill provides dry ground in a marshy area by the river.

C A ford is the easiest place to cross the river on foot.

D This is the easiest place to build a bridge over the river.

What were the advantages of building a village near a river? _______________________

____

What were the disadvantages? _______________________________________

____

Look at a map of your village, town or city. Why was it built on this site? ___________

____

The River Nile

The River Nile

The map right shows the course of the River Nile in Africa.

- Use an atlas to help you to label the following on the map:

  Red Sea

  River Nile

  Blue Nile

  White Nile

  Ethiopian Highlands

  Libyan Desert

  Sahara Desert

  Lake Nasser

  Lake Victoria

  Mount Kenya

  Mount Kilimanjaro

- Label these towns and cities:

  Cairo

  Kampala

  Alexandria

  Aswan

  Khartoum

- Label the sea that the Nile flows into.

- Label the countries that the Nile and its tributaries flow through.

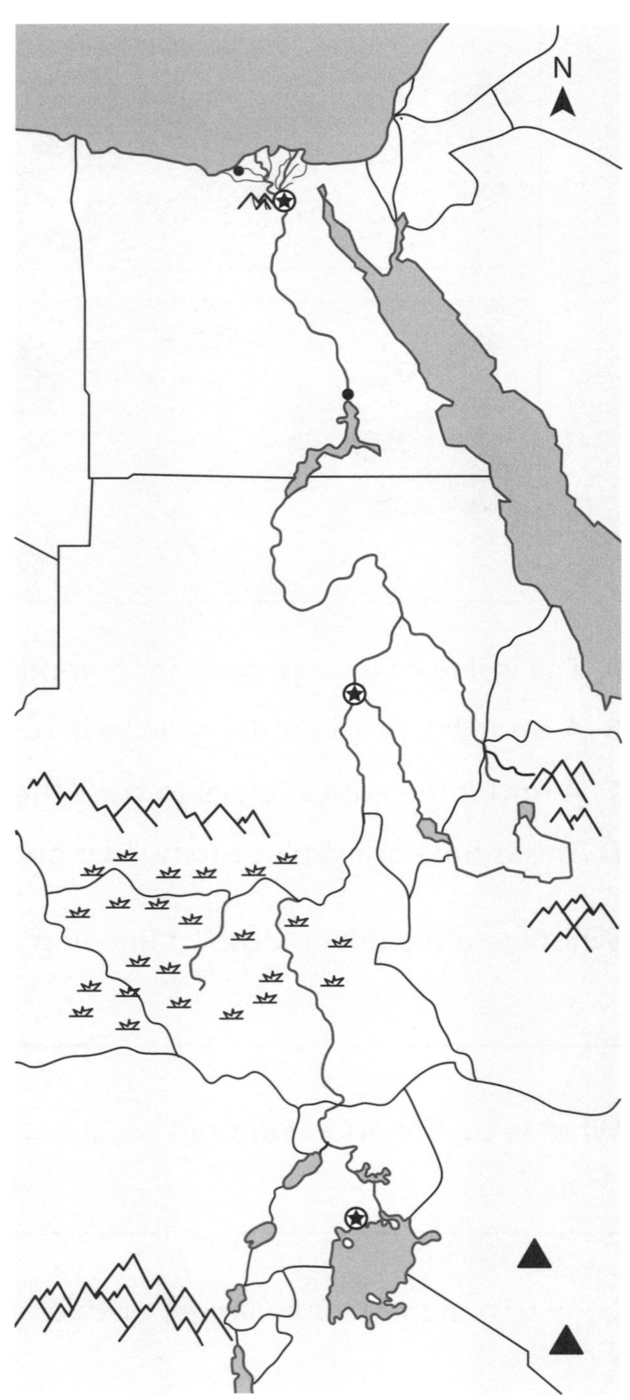

Using the Nile

Most of Egypt is a vast desert that gets little or no rain. For more than 6000 years the River Nile has enabled people to live in Egypt. Today, more than 50 million people live near the river and depend on its water.

The pictures below show some of the ways in which the people of Egypt depend on the River Nile.

1

2

3

4

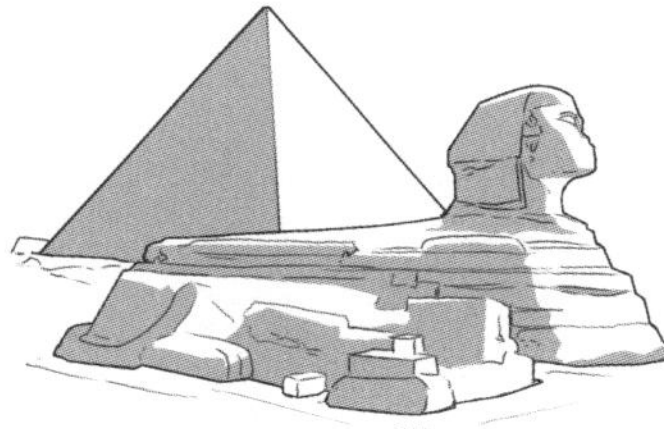

5

6

Look at the pictures carefully. Read the sentences in the box below, and match the picture numbers to the correct sentences.

- [ ] **Water from the Nile is used to irrigate crops.**
- [ ] **People catch and eat fish from the Nile.**
- [ ] **The Aswan Dam stops floods and produces electricity.**
- [ ] **Cairo grew from an army camp by the Nile.**
- [ ] **Boats carry tourists and goods along the Nile.**
- [ ] **Tourists visit the pyramids and other ancient monuments along the Nile.**

The River Amazon

The River Amazon

Here is a map showing the course of the River Amazon, from its source to its mouth.

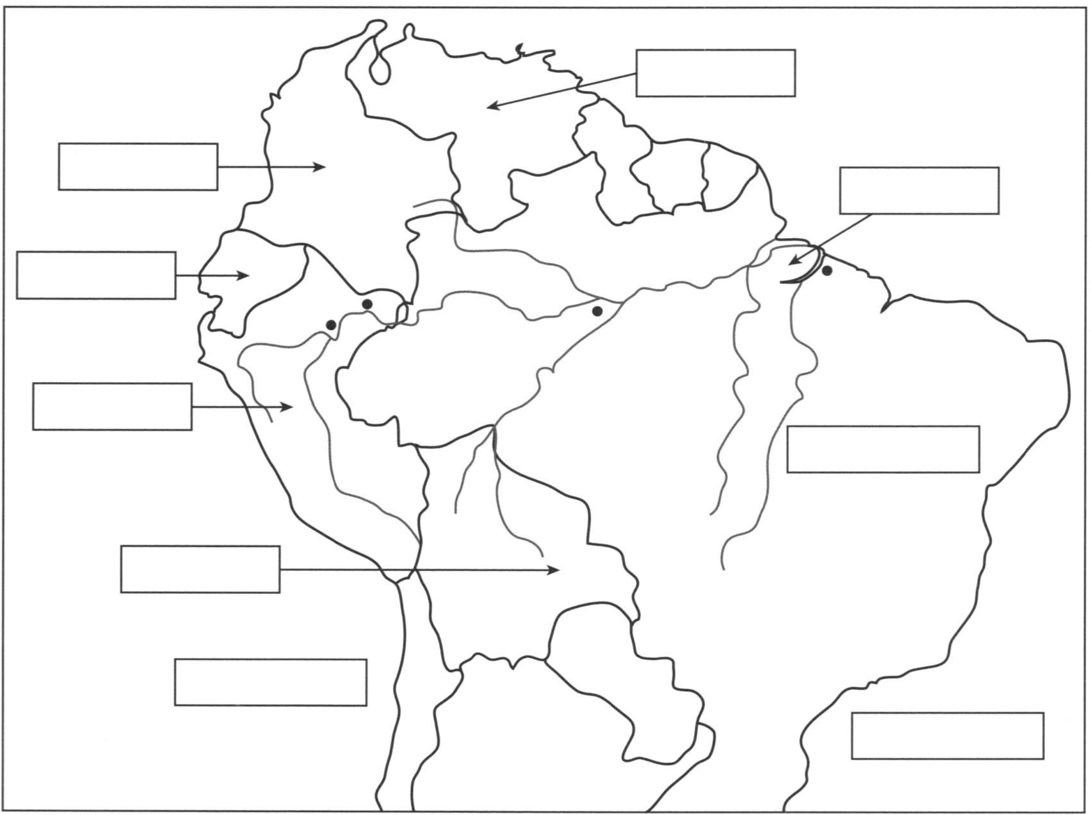

- Label the six countries shown on the map.

- Label the main River Amazon and colour it blue.

- Label the two oceans and colour them blue.

- Label the Andes mountains and colour them brown.

- Label the island at the mouth of the Amazon.

- Label the towns and cities that are marked with a black dot.

- Use reference books or the Internet to find out the populations of the towns and cities shown on the map. Write the population next to each one.

A journey down the River Amazon

Imagine you are going on a journey down the River Amazon in a dug-out canoe. You are going to travel down the Amazon from Iquitos in Peru to the city of Manaus in Brazil. Draw and label eight things you would take with you.

Now write about your journey. Include descriptions of the people and animals you might meet and any problems you might have.

The Murray River

Directions in Australia

Here is a map of Australia showing the course of the Murray River and the Darling River.

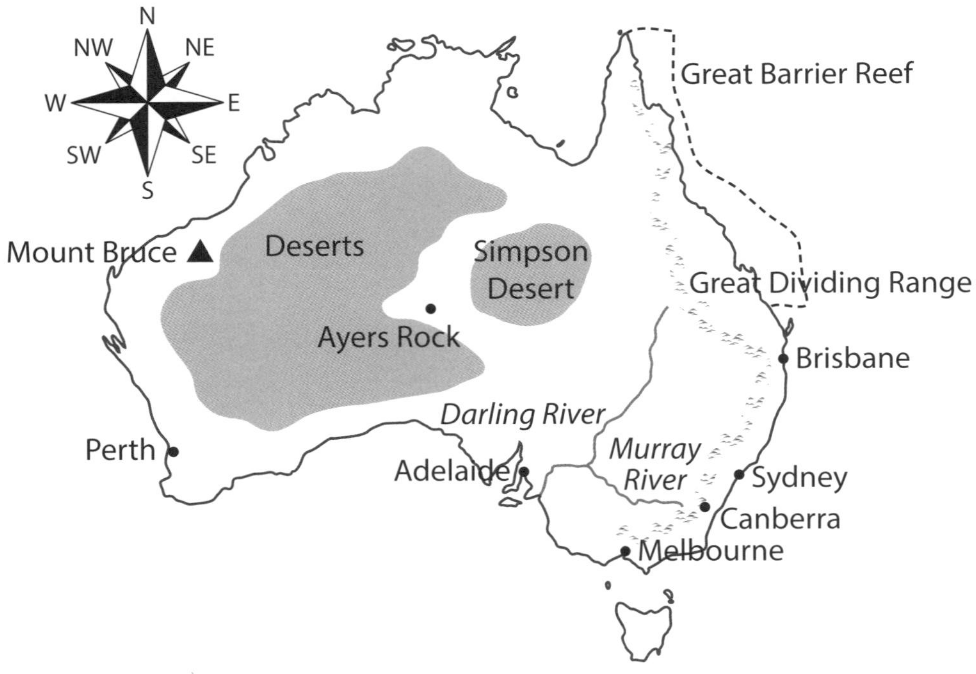

Using the map to help you, complete the following sentences:

a The Darling River flows roughly in a ________________ direction from the Great Dividing Range.

b The city of Canberra is to the ________________ of Adelaide.

c Uluru (Ayers Rock) is ________________ of Melbourne.

d The Great Barrier Reef runs along the ________________ coast of Australia.

e Mount Bruce is to the ________________ of Uluru (Ayers Rock).

f The Simpson Desert is to the ________________ of Adelaide.

g Adelaide is to the ________________ of Brisbane.

h The Great Dividing Range of mountains runs roughly parallel to the coast on

the ________________ side of Australia.

Major rivers

Use reference books and the Internet to complete the information below.

River	Continent	Length (km)	Drainage area (square km)	Countries the river passes through
Murray/Darling				
Amazon				
Nile				
Yangtze				
Mississippi/ Missouri				
Volga				

Complete these sentences:

a The longest river in the world is the _____________________. It is in the continent of

_____________________ and it flows through the countries of _____________________.

____

b The river with the largest drainage area in the world is the _____________________.

It drains an area of _____________________ square kilometres.

c The longest river in North America is the _____________________. It is

_____________________ kilometres shorter than the Nile.

d The longest river in China is the _____________________. It is also Asia's longest river.

Polluted rivers

Polluted rivers

Rivers are polluted when people dump rubbish and other waste materials
in them.

Look at this picture of a river.

Draw a cross ✕ on all the things that are polluting the river.

List the litter that you can see in and near the river. Tick a column to show
how dangerous each item is. In the last column write why you think this is.

Type of litter	Very dangerous	Slightly dangerous	Not dangerous	Reasons for my answer

Your local river or stream

Investigate your local river or stream. **ONLY** carry out this activity when an
adult is present.

Name of river or stream: ________________________   Place: ________________________

* Choose a part of the river or stream where the water is flowing fairly quickly. Does the

  water look clean or dirty? ____

* Carefully collect some of the water in a clear plastic jar and fasten the lid. Shake the jar

  thoroughly. Watch the water for five minutes. Describe what you have seen. __________

  ____

* Repeat the above activity where the water is flowing more slowly. Compare your results.

  What is it that makes the water look dirty? Where has it come from? __________________

  ____

  ____

* Find out how fast your river or stream is flowing. To do this, choose a 10-metre section of
  the river bank and mark the start and end of the section. Then find how long it takes a
  small stick, an old cork, a small orange, or other different-sized items to flow along this
  section of the river. Take several timings and find the average time for all the items.

  ____

* Repeat this activity in different parts of the river or stream. Where does it flow fastest?

  ____

* Are there any signs of pollution in your river or stream? If so, what are they? __________

  ____

Mountains, hills and maps

A relief map

Here is a relief map of the British Isles. It shows land that is over 200 metres high.

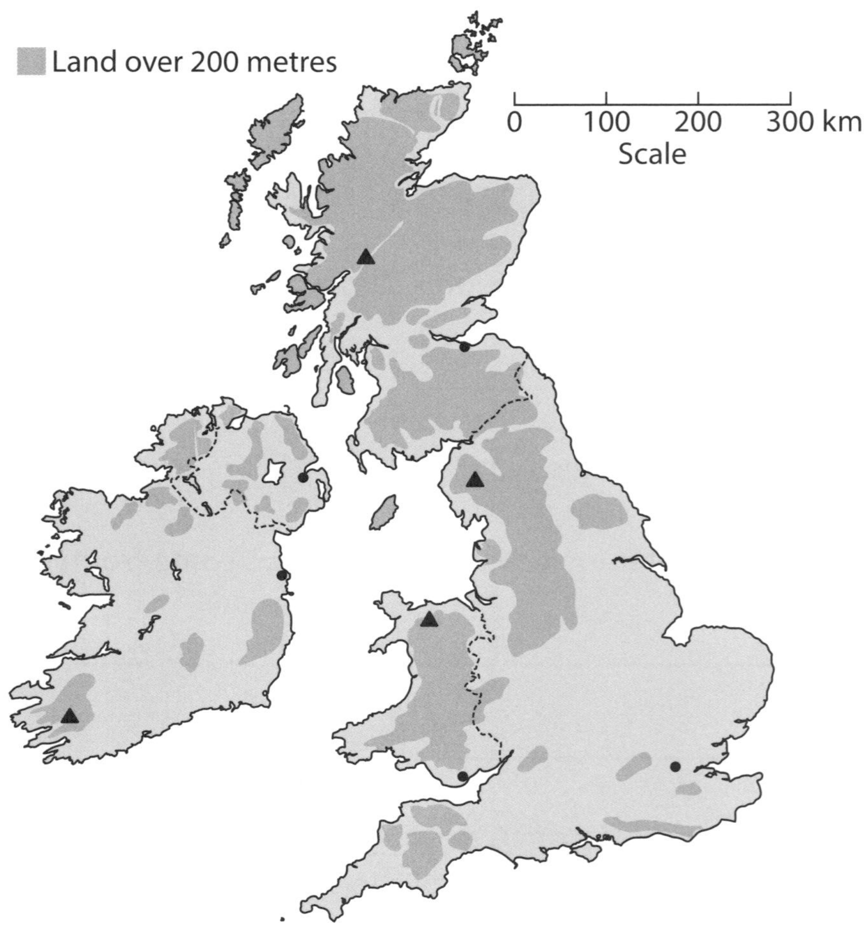

Using an atlas to help you, label the following features on your map:

- the mountains: Ben Nevis, Scafell Pike, Snowdon and Carrauntoohil

- the five capital cities: London, Edinburgh, Cardiff, Belfast and Dublin.

What do you notice about the positions of the cities? _______________________

____

 Use the Internet to help you to find out which of the areas of high land are national parks.

____

____

Contour lines

The contour lines on a map join places that are the same height above sea level. Make this model to help you to understand about contour lines.

 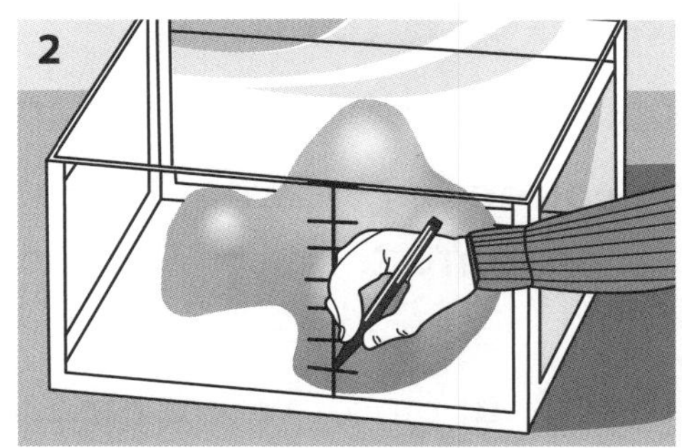

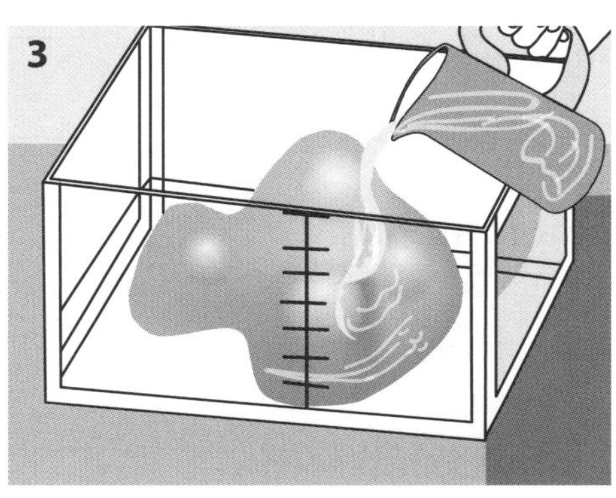 

 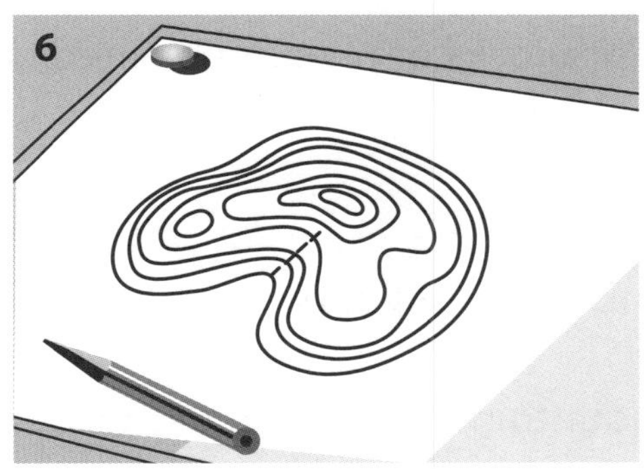

1 Make a large island using modelling clay. Make sure that it has hills and valleys.

2 Stand the model island in a large, clear plastic container. Mark a scale down the side of the container. Make sure the marks on the scale are an equal distance apart.

3 Carefully pour water into the container until it reaches the first mark on the scale. Use a waterproof marker pen to mark the line the water has reached on the island.

4 Do this several more times, marking the new water level on the island each time.

5 Carefully lift the model island out of the water. The lines you have marked on it are contour lines.

6 Look at the model from above and make a map of the island showing the contour lines.

How are mountains formed?

How mountains are formed

Here are some statements about mountains and how they are formed.

For each of the statements below, write TRUE if you think it is correct and FALSE if you think it is incorrect.

1 Most mountains form at places where the Earth's plates push towards one another with enormous force.

2 Millions of years ago Mount Everest was not a mountain, but was underneath the ocean.

3 Most mountains are found in huge rows called mountain strings.

4 There are no mountains on the bottom of the oceans.

5 As the rocks of the Earth's crust move, the cracks or breaks that form are called faults.

6 Sometimes great blocks of rock are pushed up between two faults, forming fold mountains.

7 Sometimes, as the Earth's plates push towards each other, they push the rocks up into folds called block mountains.

8 Many of the mountains that stand alone were made by volcanoes.

9 The highest mountain range on Earth is the Himalayas in Asia.

10 Mountains are built by the same forces that cause earthquakes.

11 The Alps, Andes and Himalayas are examples of block mountains.

12 There are no high mountains in Antarctica.

Volcanoes and other mountains

Some mountains were made by volcanoes. A volcano starts as a hole or crack in the Earth's crust, where the crust is weak.

The diagram below shows the insides of a volcano.

Label the diagram using words from the word box.

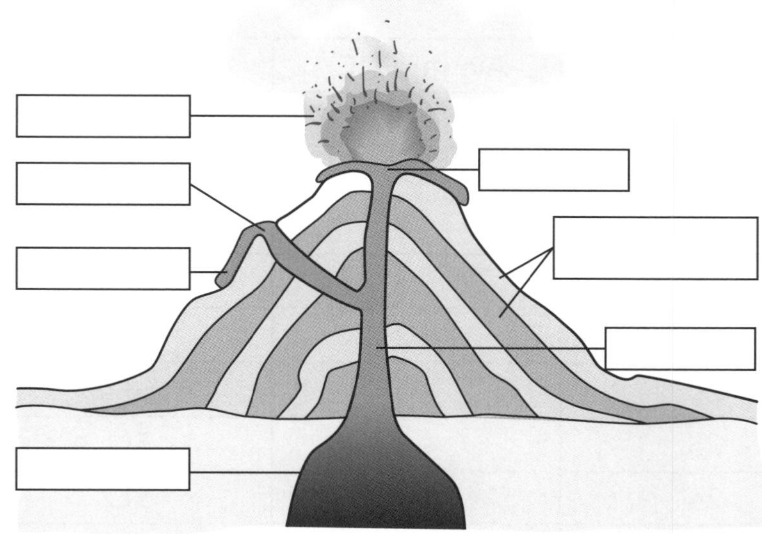

magma chamber   **secondary cone**   **crater**   **lava flow**

smoke and ash   **layers or lava and ash**   **main vent**

Use reference books and the Internet to find the names of three active volcanoes and three extinct volcanoes. For each, say which country it is found in.

Active volcanoes

a _________________________ Country _________________________

b _________________________ Country _________________________

c _________________________ Country _________________________

Extinct volcanoes

d _________________________ Country _________________________

e _________________________ Country _________________________

f _________________________ Country _________________________

Mountains and the weather

Mountain weather

Banff is a ski resort in the Rocky Mountains in Canada.

The average monthly maximum and minimum temperatures, and the average monthly rainfall and snowfall for Banff are shown below.

	Av. max (°C)	Av. min (°C)	Rainfall (mm)	Snowfall (cm)
January	−5	−15	2.4	38.2
February	0	−11	1.7	30.0
March	4	−8	1.6	27.0
April	9	−3	10.6	26.3
May	14	2	42.4	17.1
June	20	5	58.4	1.7
July	25	7	51.1	0.0
August	24	7	51.2	0.0
September	17	3	37.7	7.0
October	10	−1	15.4	18.9
November	1	−8	6.0	33.6
December	−5	−14	2.8	43.9

1 Which are the coldest three months in Banff? _______________________________

2 What is the wettest month in Banff? _______________________________

3 Which three months have the highest snowfall? _______________________________

4 In which months are the rivers likely to flood? _______________________________

____

5 When are the mountain rescue teams likely to be busiest? _______________________

6 Using the information above, how would you describe the climate of Banff?

____

Freezing and thawing rocks

On many mountains, rocks freeze and then thaw out again many times.
What effect do you think this has on the rocks? Here's how you can find out.

- Find a small piece of rock. A small piece of chalk, brick or mortar will do.

- Draw your piece of rock in box **A** below.

  Now weigh your rock, in grams. ☐ grams

- Soak the piece of rock in water overnight, then wipe it dry and weigh it again. ☐ grams

  Is it lighter or heavier than before? _______________________________________

  Why do you think this is? _______________________________________

- Wrap the piece of rock in cling film and then put it in the freezer for several days.

- Take the piece of rock out and examine it closely.

  Has it changed? _______________________________________

  If so, how? _______________________________________

- Let the rock thaw out and then put it back in the freezer for a few more days.

- Repeat this last process several more times and keep a diary of what happens to your piece of rock. Draw your rock in box **B** below.

- What happens when a piece of rock freezes and thaws many times?

____

A	B

The Himalayas

An expedition to Mount Everest

Mount Everest, in the Himalayas, is the highest mountain in the world, at 8846 metres.

- Imagine you are going to organise an expedition to climb Mount Everest.

- Write a list of the equipment and other items you would take. ________________

____

____

____

- Use an atlas to plan a route that would take you from your home to as near to

Mount Everest as possible. Write the route you would take. ________________

____

____

How far is it and in which direction would you need to travel? ________________

____

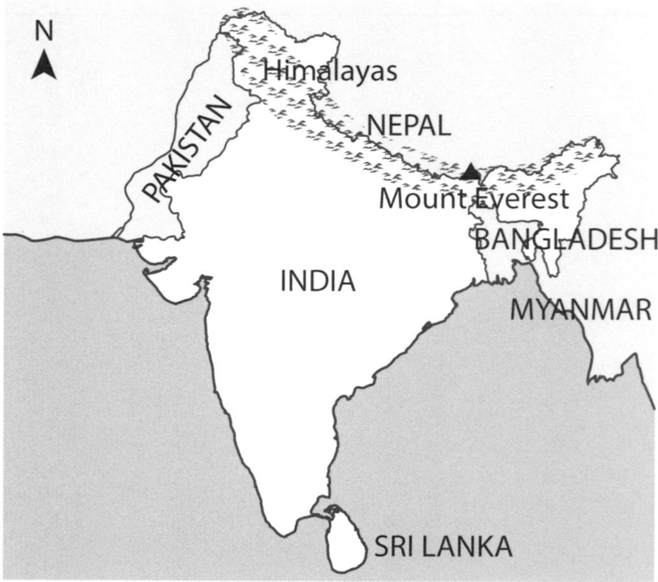

Why is litter a problem on Mount Everest?

It is very cold all the year round near the tops of very high mountains, such as Mount Everest. Like many high mountains, Mount Everest has a permanent layer of ice and snow.

Here is an experiment to show how temperature affects some items.

- Fill two clean yoghurt pots with moist soil or compost.

- Bury an apple core or a piece of banana skin, or orange or potato peeling, just under the surface of the soil in each pot. Label the pots with what you have buried and the date.

- Put one pot in a warm place, such as a sunny windowsill or near a radiator. Put the other one in a refrigerator or, if it is winter, on a windowsill outside.

- After two weeks, carefully tip out the contents of each pot onto a sheet of newspaper. What differences do you see between the two pieces of fruit or vegetable?

____

____

____

____

- Now repeat the experiment using other waste material.

Can you now explain why some litter has been lying around on high mountains,

such as Mount Everest, for many years? ______________________

____

____

Does iron rust away as quickly in a cold place as it does in a warm one? Can you find out?

World mountains

The Andes form the longest mountain range in the world. Below is a list that shows the Andes and another II very long mountain ranges. Use an atlas and the Internet to help you to complete the information.

Mountain range	Continent	Countries	Highest mountain	Height (m)
Andes				
Appalachian Mountains				
Atlas Mountains				
European Alps				
Ethiopian Highlands				
Great Dividing Range				
Pyrenees				
Himalayas				
Carpathian Mountains				
Ural Mountains				
Rocky Mountains				
Transantarctic Mountains				

Mountain heights

- Find out the heights of some mountains in several different countries.

- Draw a graph, similar to the one below, to show their heights.

The tallest mountains in the British Isles

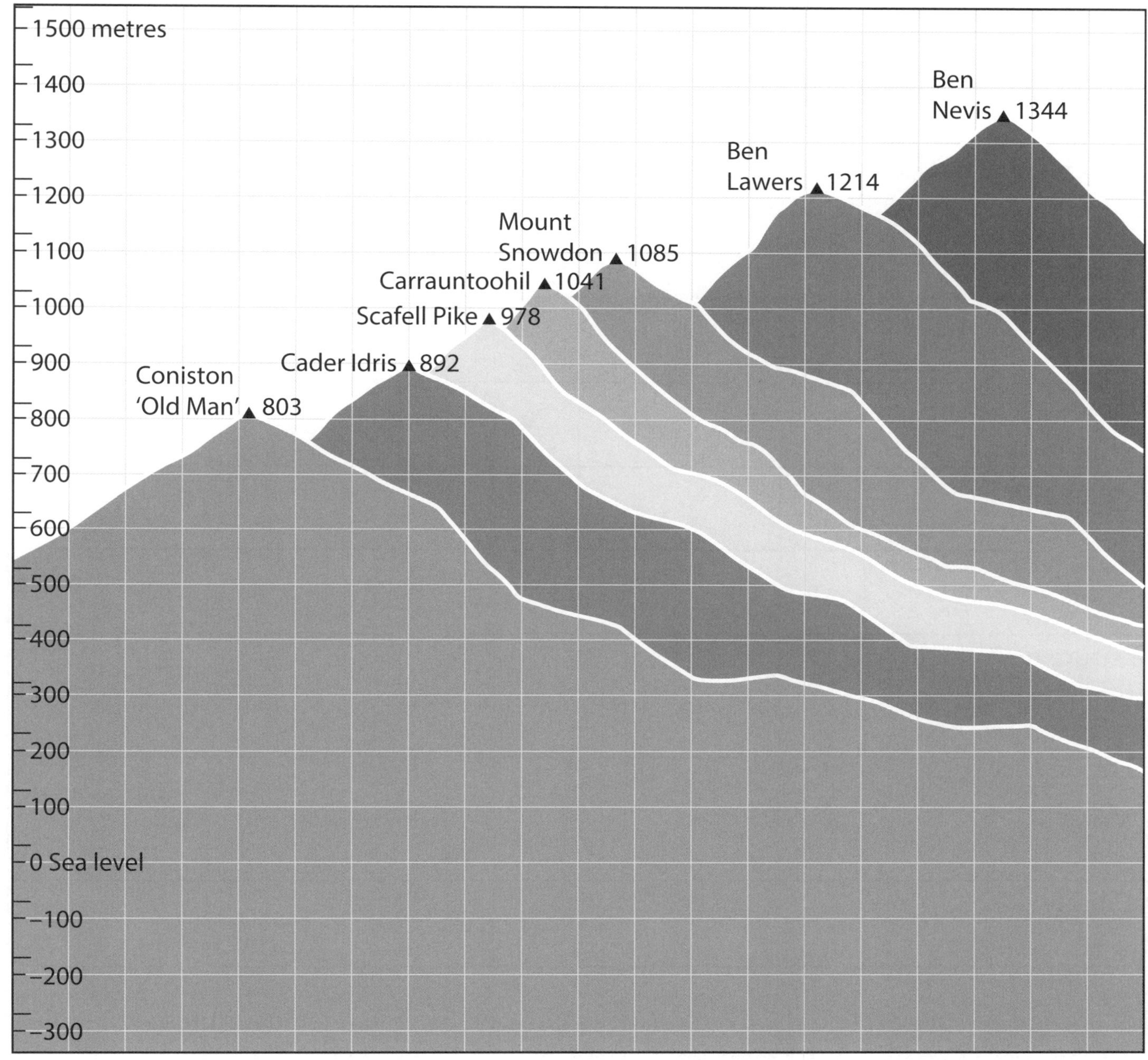

- What is the highest place in the country where you live? _______________________

- How high is it? ____

- Draw this on your graph as well.

Mountains and tourism

Mountain places

Choose three places, in three different mountain ranges, which are popular with tourists. For example, you might choose Katmandu in the Himalayas, Jasper in the Rocky Mountains in Canada and Huancayo in the Andes in Peru.

Use reference books, atlases, the Internet and holiday brochures to find out all you can about the three places. Fill in the information below.

Place	1:	2:	3:
Country			
Mountain range			
Landscape			
Work			
Leisure activities in summer			
Leisure activities in winter			
Climate			

For each of the places, find the average maximum and minimum temperatures and the average amount of precipitation (rain, hail, sleet and snow) for each month of the year.

Use graphing software on a computer to compare the climate of the three mountain places.

In what ways are they similar? ____

In what ways are they different? ____

Going to the mountains

Imagine that you are going to organise a camping holiday in the mountains for you and your family.

First, choose a mountain.

What is its name? _______________________________________

How high is it? _______________________________________

In which country is it? _______________________________________

In which mountain range is it? _______________________________________

Write a list of the clothes, equipment and other items you would take.

____

____

____

____

____

Use an atlas to plan a route that would take you from your home to as near to the mountain as possible.

Now answer these questions:

1 How far is it? _______________________________________

2 In which direction would you need to travel? _______________________________________

3 How would you travel? _______________________________________

4 Approximately how much would it cost to get there? _______________________________________

Changing coastlines

Coastal features

Look at these pictures of physical coastal features. Label each feature using words from the word box.

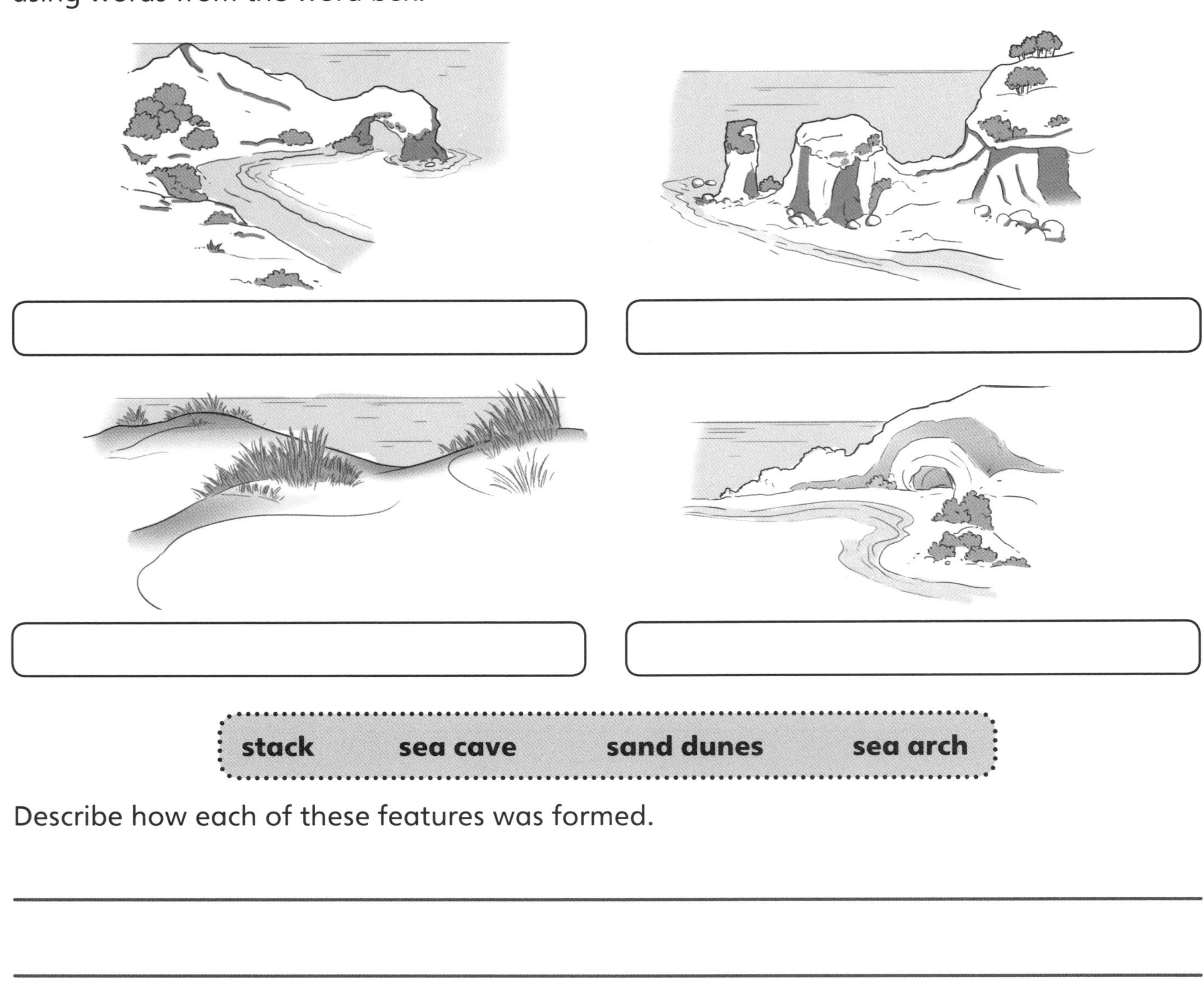

stack   sea cave   sand dunes   sea arch

Describe how each of these features was formed.

____

____

____

____

Not all coastal areas have these features. Why do you think these were formed here?

____

____

Erosion

Erosion occurs on the coast and in many other places.

- Look at each of the pictures. For each one, say what you think might happen when:

  a Farmland is ploughed down the slope. ________________________________

  ____

  b Crowds of walkers use the same mountain path. ________________________

  ____

  c Waves crash against the cliffs. ____________________________________

  ____

  d A river has created a waterfall with a plunge pool below it. ____________

  ____

  e People take a short-cut across sand dunes. ____________________________

  ____

  f A glacier slides down a mountain. ___________________________________

  ____

- Can any of these examples of erosion be prevented or repaired? ____________

  ____

Sand dunes and people

Sand dunes

On some sandy seashores, the landscape is shaped by the wind. Dry sand is blown into small hills called dunes. Eventually these are colonised by grasses and other plants. If the plant cover is damaged, strong winds can swirl the sand around into different positions.

The picture below shows some sand dunes before a gale.

Draw a picture to show how the dunes might look after a gale.

What might happen eventually to the farm in the distance if the sand dunes keep moving?

____

____

What could be done to stop this happening?

____

____

Tourist damage to sand dunes

Sand dunes are very beautiful and are very popular with holidaymakers. They are also home to rare plants and animals. However, all over the world people are damaging sand dunes. This is mainly by:

- trampling over the sand dunes

- leaving litter

- driving over the dunes in off-road vehicles

- using speed boats and water skis very near the shore.

Think about each of the problems listed above. Imagine you are in charge of an area of sand dunes. What would you do about each of the problems to protect the beauty and wildlife of the area, while still allowing people to enjoy themselves?

Trampling ____

____

____

Leaving litter ____

____

____

Driving off-road vehicles ____

____

____

Using speed boats and water skis ________________________________

____

Buildings on the coast

Physical and human features on the coast

Collect eight postcards or other pictures showing coastal scenes.

Think about the different physical and human features you can see.
Physical features are natural features such as hills, cliffs and rocks. Human features were made by people.

Use the headings below to write the information for each picture.

Picture	Name of place	Location of place	Physical features	Human features
1				
2				
3				
4				
5				
6				
7				
8				

Think about how the places are similar. Think about how they are different.

- Did you see more physical or more human features? _______________________

- Which two places have similar physical features? _______________________

- Which two places have very different physical features? _______________________

- Which two places have similar human features? _______________________

- Which two places have very different human features? _______________________

Preventing coastal erosion

The coast of East Yorkshire in north-eastern England is eroding away rapidly. On average, 2 metres of cliff are lost to the North Sea every year. During storms, the waves can destroy 10 to 20 metres of cliff at a time. Since Roman times, at least 36 villages have been lost to the sea.

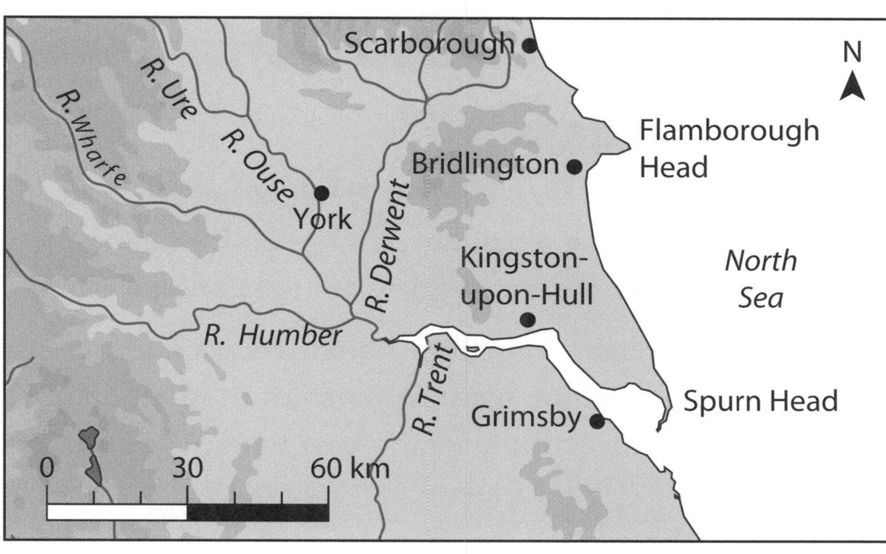

Here are some facts about the coast of East Yorkshire:

- The cliffs are made of soft clay.

- The waves usually hit the shore from the northeast.

- Sand is carried southwards by longshore drift.

Think about the coast of East Yorkshire. What would you do to protect the cliffs and coast? Draw what you would do on the map above and explain it here.

____

____

____

How and why do you think it would work?

____

____

Compare your solution with those of your friends.

Would the most expensive option be the best one to choose? _______________

Why? ____

____

Using the coast

Coastal holidays

- Imagine you are a travel agent. The four people in the pictures below want you to choose a holiday destination for them on the coast in another country. They each want to carry out their special interest or hobby.

- Use atlases, holiday brochures, guide books and the Internet to help you choose a place on the coast for each of these people. Say why you have chosen each place, and give the best time of year for each person to go on holiday.

Using the coast

Think of all the different ways in which people use the coast for work and
for leisure.

For each of these ways, decide whether they have good effects on the
environment, bad effects, or both good and bad effects.

Write the information below.

Use of the coast for work or leisure	Good effects on the environment	Bad effects on the environment

Holding back the sea

Coastal defences

People use various structures to try to stop the sea from damaging the coast and the buildings along it. These structures are called sea defences or coastal defences.

- Here are some pictures of sea defences. Use words from the word box to label each of them.

groynes	drainage pipes	breakwater	barrage or barrier	sea wall

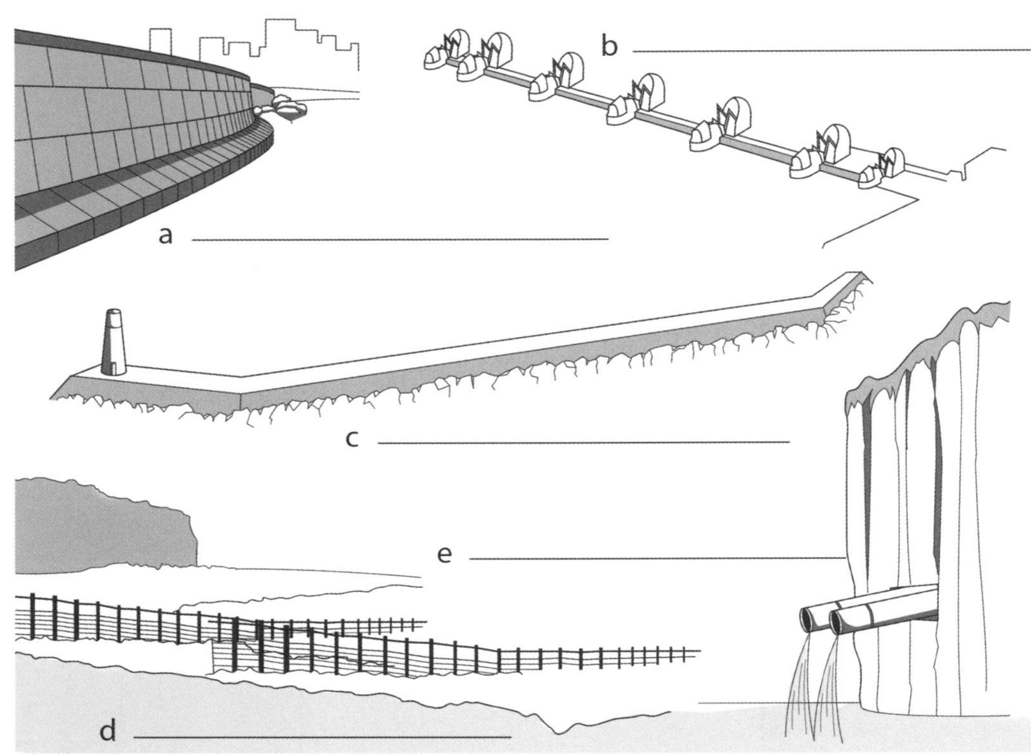

- Read the following sentences. For each of the things described, say which of the sea defences you would put in place:

a to protect the small boats in a harbour _______________________________________

b to try to stop the sand on a beach from being washed away _______________________

c to protect a low-lying village from flooding by the sea _______________________

d to stop cliffs from collapsing after heavy rain _______________________

e to stop a very high tide pushed along by the wind from flooding a city

 built on an estuary _______________________

Holding back the sea

Think about the various ways that are used to try to protect the land from flooding and being eroded by the sea. Think, in particular, about the cost of each method and what effects it has on wildlife and the environment. Complete the information below.

Sea defences	Advantages	Disadvantages
Sea walls		
Artificial reefs		
Breakwaters		
Groynes		
Draining cliffs		
Barrages		
Managed retreat		

Threats to coasts

Tourist pressures

Seaside resorts that have sandy beaches and hot, sunny weather attract many tourists. The coast of southern Spain is one example. Thousands of people go there on holiday and spend money in hotels and restaurants, or in taking trips and buying souvenirs. Many local people make their living from the tourists.

To cope with the growing number of tourists, villas, hotels and theme parks are being built. This means that fertile soil that is ideal for growing crops is rapidly becoming built upon.

- Work in a group of six. Imagine that each of you is one of the following people who live on the coast of Spain.

a builder	**a farmer**	**a fisherman**
a restaurant owner	**a shopkeeper**	**an environmentalist**

Are you **for** or **against** more tourists coming to the area? ______________________

Why? ___________________________________

____

____

____

- In your group, discuss the advantages and disadvantages of building more hotels, villas and theme parks. What did your group decide?

____

____

Threats to the coast

The picture below shows some of the threats to the coast; even parts of
the coast a long way from towns and cities.

I What evidence can you see of pollution? ________________________________

____

2 Where has this pollution come from? __________________________________

____

3 What can be done to reduce this type of pollution? ______________________

____

4 What other types of pollution could this coast face? _____________________

____

5 Are there any signs that the coast is being eroded? ______________________

____

6 What is being done to try to reduce or prevent coastal erosion? ___________

____

The growth of cities

The growth of cities

Birmingham is a city in England. It is the second largest city in England, after the capital, London.

Birmingham has a population of just over 1 million (1,085,400 to be exact).

Bucharest is the capital city of Romania.

Bucharest has a population of nearly 2 million (1,942,000 to be exact).

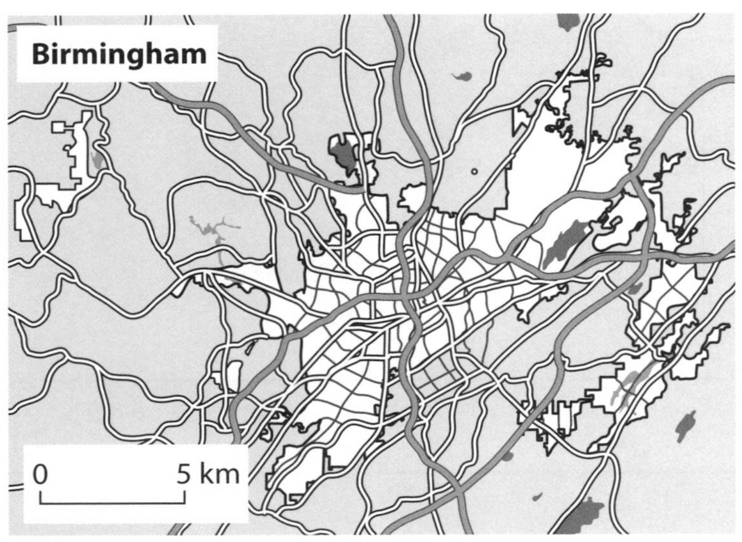

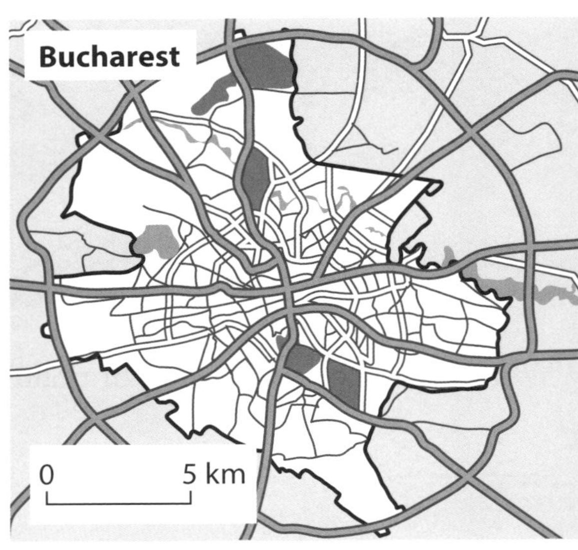

The population of Bucharest is nearly double that of Birmingham and yet these cities cover about the same area. Can you think of two possible reasons to explain this?

1 ____

2 ____

Now draw a sketch map of your own city, town or village. If possible, draw it to the same scale as the maps above. What is the population of your city, town or village?

Population: ___________________________

World cities

- On the map of the world below, write the names of the continents.

- Carefully read each clue in the boxes. Write the correct city names for each description.

a A great city on the coast of Brazil with a famous carnival.

City: _________________

b The largest city in Australia with a large harbour, which has a famous bridge over it.

City: _________________

c An ancient city on the edge of the Sahara Desert near to the pyramids.

City: _________________

d A capital city in East Africa from which many people begin safaris to see big game animals such as lions and elephants.

City: _________________

e A European city on the River Seine with a famous tower.

City: _________________

f India's second largest city, which used to be called Calcutta.

City: _________________

g A large, noisy capital city on an island off the east coast of Asia.

City: _________________

h The most populous city in the United States of America, started as a Dutch trading post on Manhattan Island in 1612.

City: _________________

i The third largest city in Canada, this is the country's chief port on the Pacific coast.

City: _________________

Kenya

- Use an atlas to find the features listed in the word box. Label them on the map below.

Equator	**Indian Ocean**	**Mount Kenya**	**Mount Elgon**	**Tana River**
Lake Turkana	**Lake Victoria**	**Tanzania**	**Nairobi**	**Mombasa**
Kisumu	**Nakuru**	**Eldoret**	**Uganda**	**South Sudan**	**Ethiopia**	**Somalia**

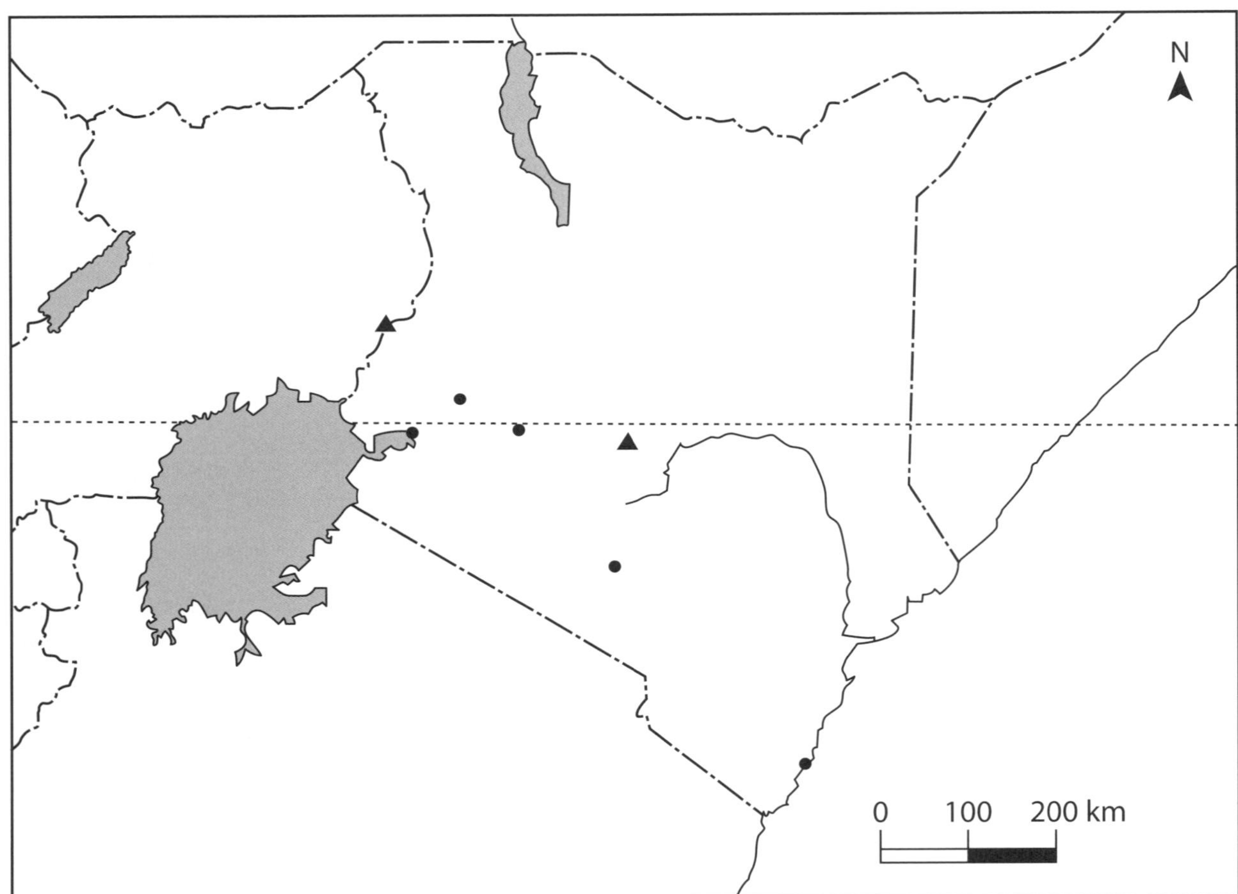

- How far is it between Nairobi and:

Mombasa? __________   Kisumu? __________   Nakuru? __________   Eldoret? __________

Going to Nairobi

Here is a map of Africa showing Kenya and Nairobi.

- Find Nairobi on a globe. Stretch a piece of thin string from where you live to Nairobi. Over which countries and oceans and seas would you fly if you took the shortest route?

____

____

- Approximately how far is it from where you live to Nairobi? _______________

- How long would it take you to fly to Nairobi? (Use airline timetables, holiday brochures

  or the Internet to help you to find out.) _______________

- How much would it cost to fly to Nairobi? _______________

Rio de Janeiro, Brazil

A postcard from Rio de Janeiro

You are going to write a postcard from Rio de Janeiro. First use these
questions to help you to find out all you can about Rio de Janeiro:

a What is the population of the city? ______________________________

b What is the climate like? ______________________________

c What famous physical features and buildings are there in the city? ______________

____

d What customs, cultural activities, sports events, etc. would attract tourists to the city?

____

e Are there any other facts you could include on your postcard? ______________

____

To ______________

On one side of the postcard, draw or paint a picture of a part of Rio
de Janeiro.

Address the other side to a friend and write what he or she would like
about Rio de Janeiro.

Finding out about Brazil

Rio de Janeiro is one of the two largest and most populated cities in Brazil.

Use an atlas, maps, books and the Internet to find out all you can about Brazil.

Write what you find out from each source of information.

	Atlases and maps	Books	Internet
Population and area			
Capital city			
Largest cities			
Currency (money)			
Main languages			
Main religions			
Name and length of the longest river			
Climate			
Farming – main crops produced			
Food (traditional dishes)			
Main industries			
Main exports			

Now answer these questions:

1   a Which source of information was easiest to use? ______________________

    b Why? ______________________

    c Which source of information was the most up to date? ______________________

2   a Did each of the sources of information give the same information in the same form?

    ____

    b If not, how were they different? ______________________

Tokyo, Japan

Tokyo and my local city

Imagine you are going to Tokyo. Find out all you can about that city ready for your visit. Compare Tokyo with a city near your home. Write what you find out.

	Tokyo	My local city
Country and continent		
Population		
Area		
Landscape		
Location in the country		
Main languages		
Currency (money)		
Houses and homes		
Famous buildings		
Other tourist attractions		
Main industries		

Collect information, each day, about the weather in Tokyo and your city.
Find a way to record the information.

Japan fact file

Tokyo is the capital city of Japan.

Find out all you can about Japan.

Fill in the fact file.

Continent	
Population	
Capital city	
Currency (money)	
Flag	
Languages spoken	
Other big cities	
Neighbouring countries	
Famous sights and interesting facts	
Trade and industry	

Food and people

Food from developing countries

Below is a list of the developed and the less-developed parts of the world.

Developed parts	Less-developed parts
Europe	Rest of Asia
Arabia	Africa
Russia	South America
Japan	
North America	
Australia	
New Zealand	

Look at the packets, jars and other containers of food in your cupboards at home and on the shelves of the local supermarket. How many foods can you find that come from:

a the developed countries of the world

b the less-developed countries of the world?

Write what you have found.

Developed countries		Less-developed countries	
Food	Country	Food	Country

How many foods came from less-developed countries? _______________________

Is it a good thing or a bad thing for less-developed countries to export food to other

countries? _______________________

  Explain your answer. _______________________

____

Food supplies

Find information about food shortages from the Internet, newspapers, and television and radio broadcasts. Colour the world map to show the places where many people do not have enough food. Colour the key.

Which continent has the most countries with food shortages? _______________

What problems are caused by lack of food? _______________

____

____

____

____

What problems are caused by eating too much food? _______________

____

____

I want, I need, I must have

Have you ever said 'I want that!', or 'I need a new pair of shoes!', or 'I must have the latest smartphone!'

The word 'want' has several meanings, but in the sentence above it means 'I would like to have that'.

The word 'need' also has several meanings, but in the sentence above it means 'I am without something I really should have; something I cannot do without'.

'I must have …' in the sentence above means 'It is important that I have …'.

Work with a group of friends and write lists of the things they say they want and the things they say they need – things they really cannot do without.

Wants	Needs

Now discuss the things you have written under the 'Needs' heading. Can your friends really not do without them? Write six things all children really do need.

____

____

Do children everywhere have the same needs? _______________________

What would be the most important need for:

a a homeless child in Syria? _______________________

b a hungry child in Eritrea? _______________________

c a sick child in Nigeria? _______________________

Rich and poor countries

Saudi Arabia and Eritrea are both desert and semi-desert countries. They are separated by the Red Sea. Saudi Arabia is a large, rich country that gains much of its wealth from exporting oil to other countries. Eritrea is a small, poor country that has suffered disastrous droughts, famines and many years of civil war.

Carefully read the pairs of facts below. If you think the fact refers to Saudi Arabia, write **SA** against it. If you think the fact refers to Eritrea, write **E** against it. The first one has been done for you.

Capital city

Riyadh — SA

Asmara — E

Area

121,320 square kilometres

1,960,582 square kilometres

Population

28,829,000

6,234,000

Annual income per person

(US dollars) 25,852

(US dollars) 544

Life expectancy

57 years

74 years

Main exports

Oil and natural gas, wheat and dates

Animal hides, textiles, salt, cement, citrus fruit

Number of primary schools

549

15,123

Number of doctors

40,360

3117

Number of motor vehicles

68,574

9,686,544

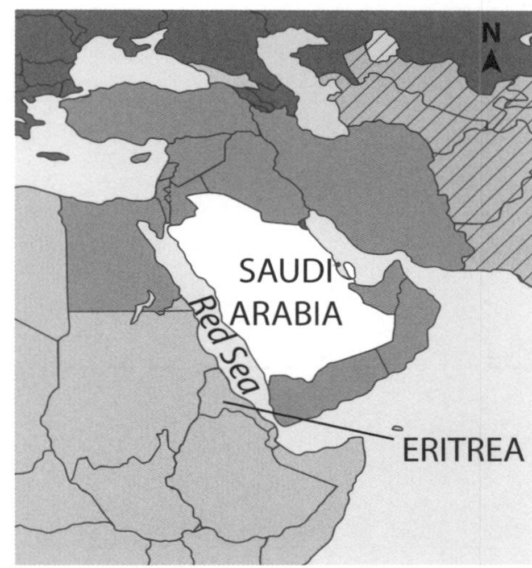

Famines caused by nature

Droughts, famine and refugees

Watch the television news and weather reports, and study the newspapers and the Internet to find information about famines caused by nature.

Record any details of droughts, famines or refugees. Fill in a box for each one saying what happened.

From each box, draw a line to the correct country to say where it happened.

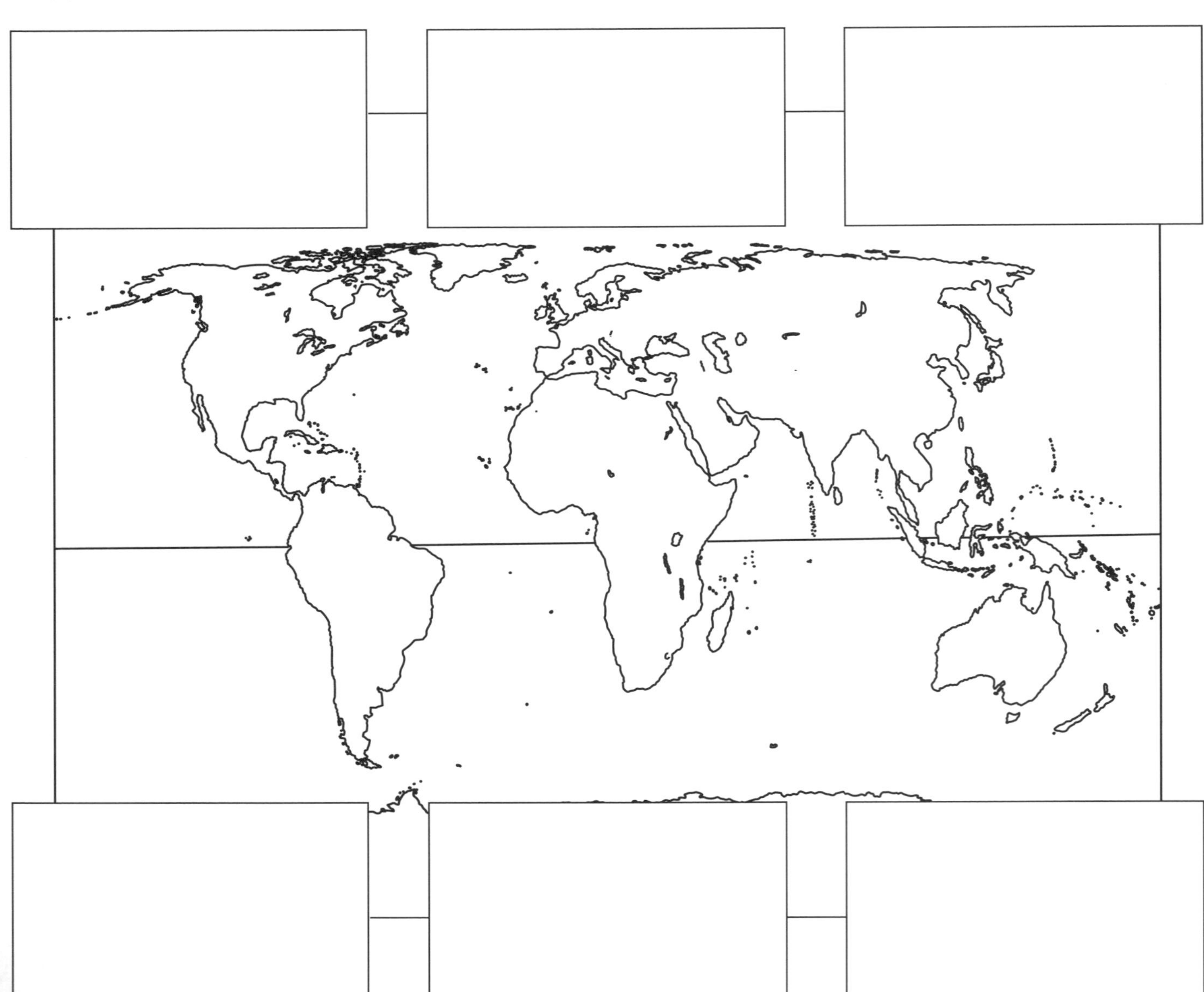

Refugees and migrants

In poorer countries, people wanting a better standard of living often move from the countryside to cities.

If people choose to move to another country it is called international migration. This kind of migration is voluntary.

There are times when people have to move, either to another part of their country or to a new country. They may have suffered because of natural disasters, ill treatment, famine or war. This forced migration makes people become refugees.

- Look at the drawings below. Read what the migrant people say.

- Write in the empty boxes whether their migration was:
  a country to city (C) or international (I); **b** voluntary (V) or forced (F)

A

B

C

D

E

F

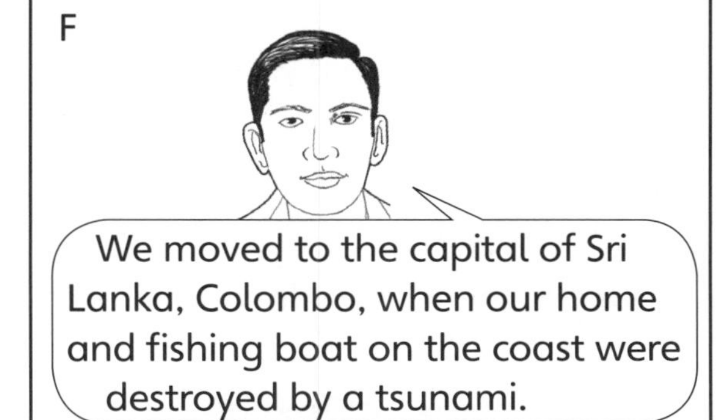

Famines caused by people

The North-South divide

All countries have some poor people, but there are many more poor people in the less-developed countries of the South than there are in the developed countries of the North.

The map below shows the North-South divide.

The information below shows some of the important differences between the developed countries of the North and the less-developed countries of the South.

	North	South
Proportion of the world's population	25%	75%
Average life expectancy	70 years	50 years
Proportion of the world's income	80%	20%
Proportion of the world's industry	90%	10%

Draw a bar chart or pie chart to show the data above.

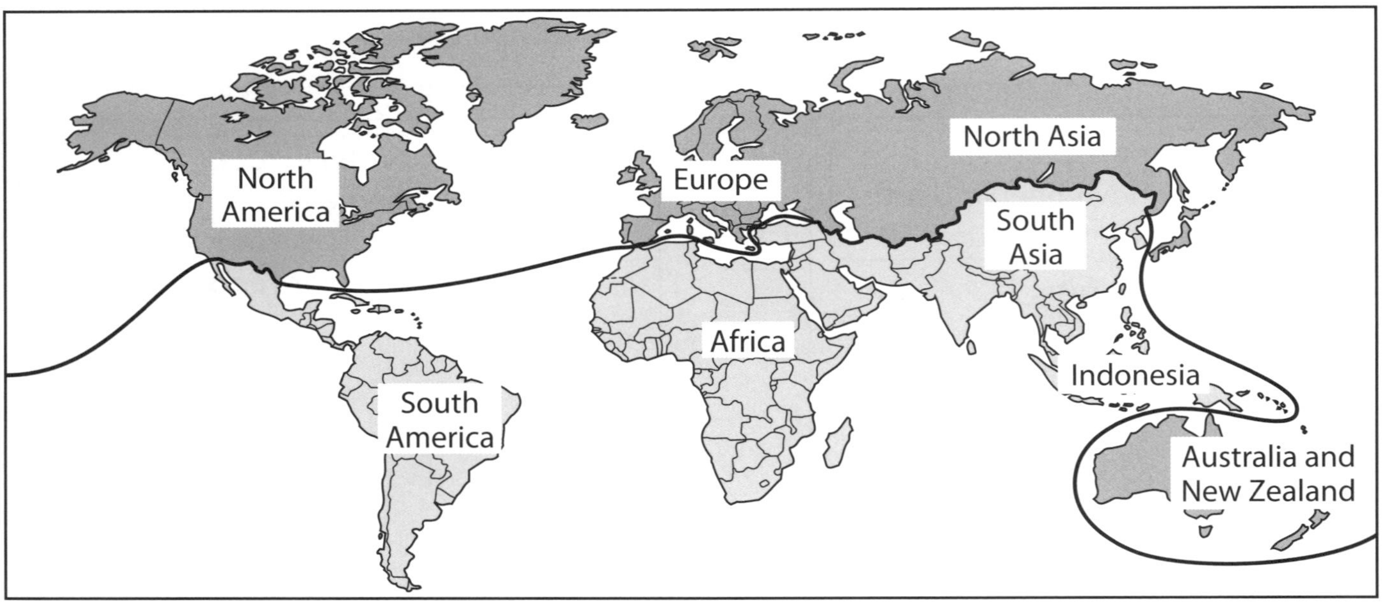

From which of the countries in the South do we obtain some of our food?

In which part of the world, North or South, do famines usually occur? ___________

Discuss with a friend what we can do to make the world a fairer place.

Why do famines still occur?

There is more than enough food for everyone in the world and yet famine still exists.

What is famine? ____

____

Write three natural disasters that can lead to famine in the 21st century.

1 ____

2 ____

3 ____

Write three human causes of famine.

1 ____

2 ____

3 ____

Which one cause of famine do you think is most difficult to solve? ___________

Why? ____

____

If you were head of the government, how would you solve the problem of famine? ______

____

____

Is there anything you can do in your life to help prevent famine? ____________

____

How might this help? ____

____

Map of the World

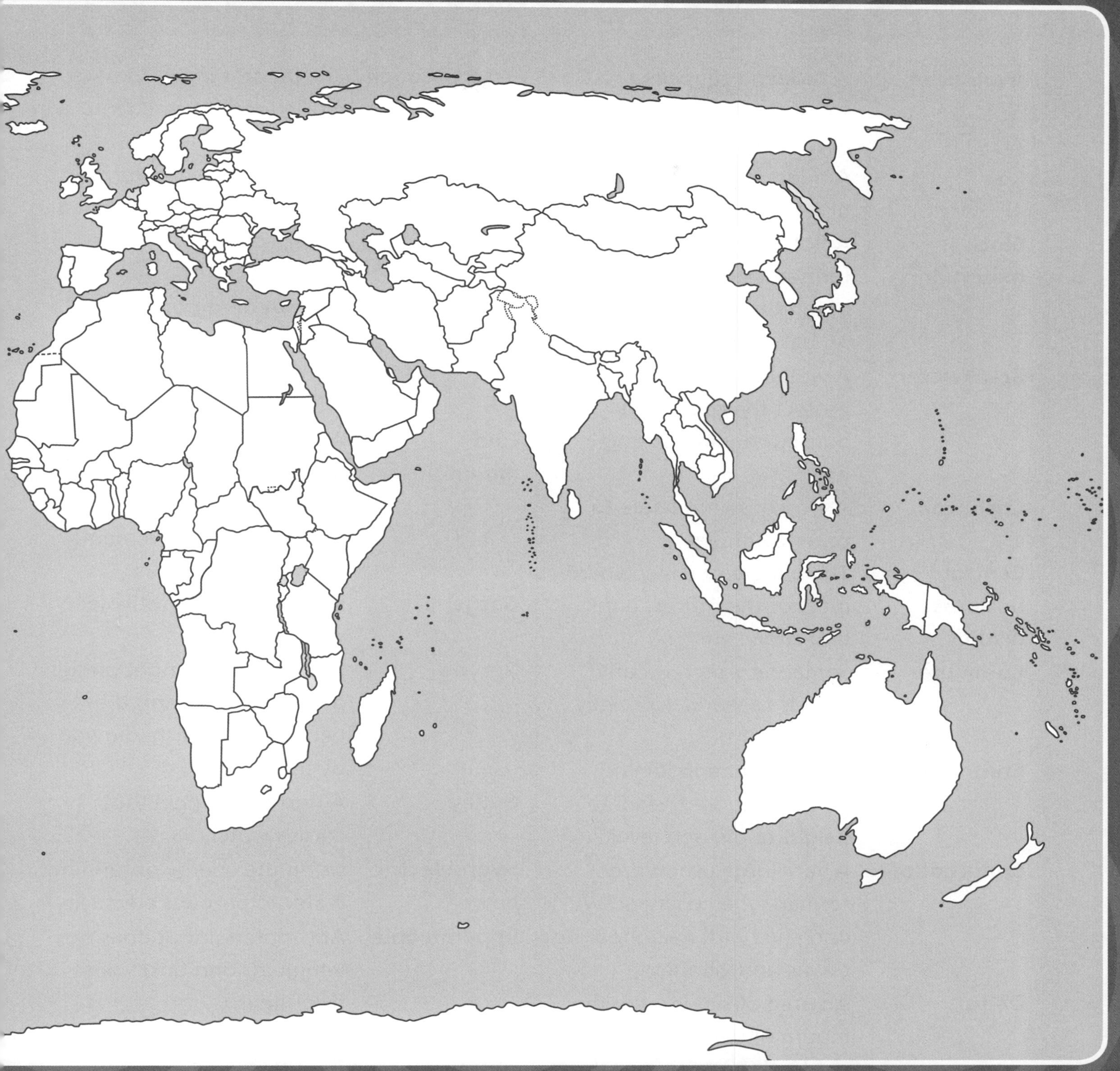

Glossary

Avalanche	A sudden collapse of snow down the side of a mountain.
Bar	A ridge of sand or shingle across a bay or river mouth.
Block mountain	A flat-topped mountain formed when a large block of land has been pushed up between two faults.
Breakwater	A wall built into the sea to protect the coast or a harbour against strong waves.
Cash crop	A crop grown for sale to other countries.
Central business district	The middle of a city where most of the offices, banks and shops are.
Commuter	Someone who regularly travels to work, especially by train or bus.
Contour line	A line on a map joining points that are the same height above sea level.
Conurbation	A very large urban area formed when a city grows and joins with neighbouring towns and villages.
Delta	An area of flat land at the mouth of a river, which is made of mud dumped there by the river. Many deltas are shaped like a triangle.
Deposit	Dropping mud, sand and pebbles when moving water is slowed down.
Desertification	The process in which once fertile land is turned into desert.
Evaporate	When liquid water turns into the invisible gas called water vapour, it is said to evaporate.
Floodplain	The flat area bordering a river formed from the mud deposited by the river when it floods.
Fold mountain	A mountain that is formed by folds or ridges of the Earth's crust that are pushed up by movements of the Earth's plates.
Gorge	A narrow valley with steep sides.
Groyne	A fence built at right angles to the shore to stop the beach being washed away by the sea.
Gulley	A narrow channel that carries water.
Hydroelectric power	Using the energy of running water to produce electricity.
Impermeable	Any material that does not let liquids and gases pass through it.
Lagoon	A shallow lake separated from the sea by a sandbank or spit.
Leeward	Facing away from the wind.
Longshore drift	The slow movement of pieces of rock along a beach when waves move towards the shore at an angle.

Malnourished	Someone who does not have enough food or the right kinds of food to eat.
Meander	A large S-shaped bend in a river.
Mountain	An area of high ground that is at least 300 metres higher than the ground around it.
National park	A large area of land set aside so that its beautiful scenery is not spoiled and so that its plants and animals can be protected.
Obese	Someone who is very fat.
Oxbow lake	A lake made when a river changes course and cuts off a meander.
Permeable	Any material that lets liquids and gases pass through it.
Port	A place where ships can be loaded and unloaded.
Rain shadow	The sheltered side of a mountain where there is less rainfall than on the other, windward, side.
Range	A row or line of mountains.
Relief map	A map that shows how high the different parts of a region, country or continent are.
Reservoir	A large artificial lake used to store water for drinking, irrigating crops, producing electricity or to prevent a river flooding.
Scale	A way of showing distances on a map.
Scree	The pieces of rock that collect at the bottom of a steep mountain slope.
Sea wall	A wall made of concrete or stone built to protect beaches and buildings from the sea.
Snow line	The line on a mountain above which it is so cold that snow covers the ground, even in summer.
Source	The place where a river rises.
Spit	A ridge of sand or shingle joined to the land at one end and sticking out into the sea at the other.
Suburb	An area on the outskirts of a town or city where many people live.
Topographic map	A map showing both natural features, such as hills, mountains, lakes and rivers, as well as human features, such as roads, railways and towns.
Trawler	A fishing boat that pulls a large net behind it.
Valley	An area of low land between hills or mountains.
Water cycle	The movement of water from the oceans, seas and other wet surfaces to the air, then back to the ground, oceans and seas again.
Weathering	The breaking up of rocks by heat, cold, ice and rainwater.
Windward	The side that faces the wind.